Garden landscape Detail Design - 1

庭院细部元素设计

路面、坡道、台阶、围墙

①

中国林业出版社
China Forestry Publishing House

图书在版编目（CIP）数据

庭院细部元素设计.①，路面、坡道、台阶、围墙/《庭院细部元素设计》编委会编.
-- 北京：中国林业出版社，2016.5

ISBN 978-7-5038-8500-6

Ⅰ.①庭… Ⅱ.①庭… Ⅲ.①庭院-景观设计 Ⅳ.① TU986.4

中国版本图书馆CIP数据核字(2016)第078636号

《庭院细部元素设计》编委会

◎ 编委会成员名单

主　　编：董君
编写成员：董君　李晓娟　曾勇　梁怡婷　贾濛　李通宇　姚美慧
　　　　　刘丹　张欣　钱瑾　翟继祥　王与娟　李艳君　温国兴
　　　　　黄京娜　罗国华　夏茜　张敏　滕德会　周英桂　李伟进

◎ 特别鸣谢：北京吉典博图文化传播有限公司

中国林业出版社 · 建筑与家居出版中心

责任编辑：纪亮　王思源
联系电话：010-8314 3518

出版：中国林业出版社
　　　（100009 北京西城区德内大街刘海胡同7号）
http://lycb.forestry.gov.cn/
E-mail：cfphz@public.bta.net.cn
电话：（010）8314 3518
发行：中国林业出版社
印刷：北京利丰雅高长城印刷有限公司
版次：2016年6月第1版
印次：2016年6月第1次
开本：235mm×235mm 1/12
印张：16
字数：200千字
定价：99.00元

鸣谢
因稿件繁多内容多样，书中部分作品无法及时联系到作者，请作者通过编辑部与主编联系获取样书，并在此表示感谢。

目录 1
CONTENTS

路面　5~192

坡道　193~208

台阶　209~276

围墙　277~320

GARDEN LANDSCAPE DETAIL DESIGN · PAVEMENT
路面

　　庭院的意义在于居住者对它所提供的生活方式的认同，在于它所带来的院落"能指"（居住环境）和"所指"（居住者的生活方式、社会交往方式）的和谐关系，而此代码在现代集合住宅中的被扭曲和遗忘引发了居民对院落中亲密的邻里关系的怀念。

　　庭院作为人为的自然空间，它将自然和人工相结合起来，让人产生一种回归自然的向往。庭院中往往常采用小巧、精致的手法，设置植物、院路、亭廊、假山、雕塑、水池等，既表现出个性化的特点，同时给人们带来自然美、人工美的艺术享受。

Agnatus eiumendae voluptamusa custet et demporeribus minverchil mi, unte magnit eum et que doloreh enihili quaspe vendi ium nume sum vita doloria nosam as utendi te solori beribus pa consequi idias maiorem porestiis exerendias as es ut volupit, el earum labo. Itatemp orerenes ipsunt, simusap idustenet explacea et.
con rat doluptatus arum et que plit occum et iunt quam, voluptia parcius dollect aturehentum abo. Neque voluptas seque sam, qui opta volorib ernate lat latem eumet la quis qui ut arit, am reperates dolore.

1-3.北京市 - 竹溪园
4.PA-323-05
5-7.Private Residence, San Francisco, California
8.Tables of Water, Lake Washington, Washingto
9.傍花村温泉度假村

Pavement ○ 路面

Pavement ○ 路面

1,2.Aley, Lebanon Date of Completion-消失点
3.Algiers, Algeria Area-Douyna公园
4.上海市-大华锦绣
5.泉州-奥林匹克花园别墅区
6,7.上海-别墅花园

Pavement ○ 路面

1.以色列贝尔谢巴-BGU University Entrance Square & Art Gallery
2.新加坡-City Square Urban Park
3.威尼斯
4.佘山 - 东紫园
5.加拿大蒙特利尔-Square Dorchester-Place du Canada
6.罗德岛普罗维登斯 - Dunkin Donuts Plaza - Horizon Garden
7.悉尼 - 海洋生物站公园

1.4 佘山的梦想87A0918
2.533-03
3-6.Aalborg, Denmark-奥尔堡滨水 - 港口与城市连接
7.Adelaide Zoo Giant Panda Forest

Pavement ○ 路面

Pavement ○ 路面

1-4.Adelaide Zoo Giant Panda Forest
5.Alameda, USA Date of Completion-海湾船艇公司

1.Anchorage, Alaska, USA-Anchorage Museum Expansion
2.Barcelona, Spain Date of Completion-Tanatorio Ronda de Dalt花园
3,4.Australia-Adelaide Zoo Entrance Precinct
5.Austin, USA Date of Completion-串联式溪边住宅
6-9.Australia-Adelaide Zoo Entrance Precinct

Pavement ○ 路面

1.Barcelona-蒙锥克公园的新缆车车站-Cims de Montjuic_Montjuic Heights_11
2.Bassil Mountain Escape, Faqra, Lebanon
3,4.Belo Horizonte, Brazil Date of Completion-Pampulha广场
5,6.Berlin-VISITOR
7,8.Berlin--三角铁路站公园--Com_Loidl_BerlinPARK_20110904-0667

Pavement ○ 路面

Pavement ○ 路面

1,2.Berlin－三角铁路站公园
3,4.Cairnlea, Victoria-Cairnlea
5.Canberra, Australia Date of Completion-澳大利亚国家美术馆-澳大利亚花园
6.Castell D'emporda露天酒店
7.Brumadinho, Brazil Date of Completion-Burle Marx教育中心

1,2.Canada-cSimcoe WaveDeck
3.Connecticut Country House, Westport, Connecticut
4.Castell D'emporda露天酒店
5-7.Charleston, USA Date of Completion-查尔斯顿滨水公园 (2).tif

Pavement ○ 路面

Pavement ○ 路面

1-3.Corner of Hay and William Streets, Perth, WA-Wesley Quarter
4.Expo 2008 Main Building
5.Corner of Hay and William Streets, Perth, WA-Wesley Quarter
6，7.Erman Residence, San Francisco, California

1. Grünewald公共果园
2，3. Groningen – 马提尼医院
4-6. Grünewald公共果园
7. Greenwich Residence, Greenwich, Connecticut
8. Hilltop Residence, Seattle, WA

Pavement ○ 路面

Pavement ○ 路面

1. Hilltop Residence, Seattle, WA
2. Holstebro, Denmark—If it is outstanding, celebrate it twice
3. Expo 2008 Main Building
4. Fredericia – Temporary Park
5. Faqra－魔力居所FEJ_9071

1. Houston, TX–The Brochstein Pavilion
2. Lakeway Drive – In Redevlopment
3, 4. Il nuovo Parco di Jean Nouvel a Barcellona
5. Holstebro, Denmark–If it is outstanding, celebrate it twice
6, 7. House by the Creek, Dallas, Texas
8. Houston, TX–The Brochstein Pavilion

Pavement ○ 路面

1-3.Lakeway Drive – In Redevlopment
4.Lee Landscape, Calistoga, CA
5.Madrid-欧·德尼尔街疗养院花园
6.Madrid RIO
7.Malibu Beach House, Malibu, California

Pavement ○ 路面

1-3.Tables of Water, Lake Washington, Washington
4.Taiwan, China Date of Completion-日月潭观光局管理处
5,6.Los Angeles, CA-Wilmington Waterfront Park
7.Lunada Bay Residence, Palos Verdes Peninsula, Southern Californi
8.Madrid RIO

Pavement ○ 路面

1,2.Malinalco House, Malinalco, State of Mexico, Mexico
3.Manhattan Roof Terrace, New York, New York
4.Melboume－圣基尔达海滨城的交通线
5-7.New York-Beach House, Amagansett

Pavement ○ 路面

1.PA-323-03
2,3.Padaro Lane
4.PA-323-014
5.Plenty Road, Bundoora, Victoria-University Hill Town Centre && Wetland
6,7.Pamet Valley

Pavement ○ 路面

Pavement ○ 路面

1，2.New York-Beach House, Amagansett
3，4.New York-南部河畔公园
5，6.Oss, the Netherlands Date of Completion-库肯霍夫街区

41

Pavement ○ 路面

1.Rotterdam－医院办公楼景观－Maasstad Hospital 04－seating elements by day
2.RGA Landscape Architects www.rga-pd.com
3.Salvokop, Pretoria, Gauteng–The Freedom Park
4.San Luis Potosí, Mexico Date of Completion－圣路易总体规划
5.Quartz Mountain Residence, Paradise Valley, Arizona

1，2.San Luis Potosí, Mexico Date of Completion-
圣路易总体规划
3.Shreveport－雨花园别墅
4，5.Santa Monica－欧几里德公园－07 RCH_Euclid Park_Bonner_5481_009C
6，7.Seattle, Washington－奥林匹克雕塑公园

Pavement ○ 路面

Pavement ○ 路面

1.Seattle, Washington-奥林匹克雕塑公园
2-5.Stanislaw Lem memorial - Garden of Experience - Educational Park in Krakow
6.Sitges-肯·罗伯特公园07 vall central
7，9.Sonoma Vineyard, Glen Ellen, California
8，10.South Africa-University of Johannesburg Arts Centre

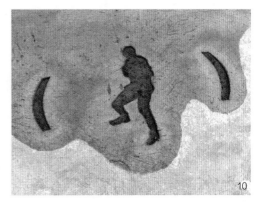

Pavement ○ 路面

1，3.阿姆斯特丹-辛克尔小岛
2.阿肯色州小石城-Little Rock Courthouse
4.埃布罗河畔
5.Wittock Residence
6.Toms River, New Jersey-Ocean County Public Library – BarCODE Luminescence
7.Urban Play Garden
8.Toronto, Ontario, Canada-HtO

1. Xochimilco, Mexico City.-XOCHIMILCO ECOLOGICAL PARK
2. Villengarten Krantz
3. 埃布罗河畔
4. Zapopan-七庭院住宅＝7 Patios House
5，6. Woody Creek Garden, Pitkin County, Colorado

1，2.埃布罗河畔
3，4.爱尔兰都柏林-Cleansweep
5-6.爱尔兰-"自我"圣坛

Pavement ○ 路面

1,2.Toronto, Ontario, Canada-HtO
3.Unfolding Terrace, Dumbo, Brooklyn, New York
4.Villa H. St. Gilgen
5,6.爱尔兰都柏林-Cleansweep
7.爱尔兰都柏林-凡丹戈
8.Lima-绿意来袭

Pavement ○ 路面

55

Pavement ○ 路面

1.爱莫利维尔码头
2.爱涛漪水园承泽苑16栋
3.爱尔兰都柏林-红杉（展示花园）
4-6.爱涛漪水园承泽苑16栋

1-3.爱涛漪水园承泽苑16栋
4.澳大利亚悉尼-Foley公园
5,6.安徽省-合肥政务文化主题公园
7,8.安斯伯利镇区
9.盎格滨海路

Pavement ○ 路面

Pavement 路面

1-3.奥格斯堡Wollmarthof住宅
4-6.澳大利亚-不莱梅洲际中学
7, 8.澳大利亚堪培拉国家紧急救灾服务纪念馆

1. 澳大利亚-库吉港
2. 澳大利亚墨尔本-墨尔本717号Bourke大街
3-8. 澳大利亚墨尔本-墨尔本国际会议会展中心

Pavement ○ 路面

1,3.德国巴伐利亚州-伯格豪森的巴伐利亚州园林展 - 城市公园
2.德国柏林-国家展会（ULAP）广场
4,5.澳大利亚悉尼-Bondi 至Bronte 滨海走廊
6,7.澳大利亚悉尼-Foley公园

Pavement ○ 路面

Pavement 路面

1. 澳大利亚悉尼-Victoria Park Public Domain
2，3.芭堤雅希尔顿酒店
4.澳大利亚悉尼-悉尼5号湿地0
5.百家湖
6.白香果

Pavement ○ 路面

1-3.柏林-Südkreuz火车站
4.鲍恩海滨
5-7.傍花村温泉度假村

Pavement 路面

1. 鲍恩海滨
2-5. 北京燕西台
6. 北京-八达岭下的美式小院
7. 北京-波特兰
8. 北京-翠湖——教授的乐园

1.北京-低碳住家——北京褐石公寓改造设计
2.北京-翠湖——教授的乐园
3-7.北京-过山车

Pavement ○ 路面

Pavement ○ 路面

1. 北京-花石间
2. 北京-京都高尔夫中式别墅
3. 北京-龙湾ZR8T3052
4. 北京-纳帕尔湾
5. 北京-龙湾别墅和院
6. 北京-润泽庄园
7. 北京市-竹溪园

1.澳大利亚悉尼-Foley公园
2.北京市-西山美墅馆
3,4.北京市-竹溪园
5.百家湖
6,7.柏林-Südkreuz火车站
8.澳大利亚悉尼-Pimelea Play Grounds Western Sydney Parklands
9.澳大利亚悉尼-Victoria Park Public Domain
10.澳大利亚悉尼-Ng别墅花园

Pavement ○ 路面

1-3，5.北京-泰禾运河岸上的院子
4.北京市-润泽庄园2
6.北京-天竺新新家园
7.北京-新新家园
8.北京-易郡别墅
9.北京-燕西台

Pavement ○ 路面

Pavement ○ 路面

1，2.北京-易郡别墅
3.碧湖山庄
4.北京圆明园-左右间咖啡的院
5.泉州－奥林匹克花园别墅区
6.北悉尼地区-BP公司遗址公园
7.比利时Kapellen-Meeting square Kapellen

1.滨海湾花园群
2-4.波特兰花园60号
5,7.布卡克海滩
6.波士顿－帕梅特谷

Pavement ○ 路面

1-4.财富公馆
5.成都美景金山
6.成都-翡翠城汇锦云天1
7.成都双流-和贵馨城

Pavement ○ 路面

Pavement ○ 路面

1，2.比利时Kapellen-Meeting square Kapellen
3.达尔文海滨公共区域
4.成都双流-和贵馨城
5.碧湖别墅
6.成都-紫檀山
7.成都-玉都别墅庭园

1-5.达克兰的维多利亚港
6.大豪山林21号
7,8.大湖山庄

Pavement ○ 路面

1-3.大湖山庄
4-7.袋鼠湾
8.丹麦,奥尔堡-DANISH CROWN-AREAS, NR. SUNDBY BRYGGE

Pavement ○ 路面

Pavement ○ 路面

1-3.丹麦，奥尔堡-DANISH CROWN-AREAS, NR. SUNDBY BRYGGE
4-6.丹麦-哥本哈根的西北公园

1-4.丹麦哥本哈根-云
5.德国 柏林-南"制革厂"岛

Pavement ○ 路面

1-3.德国巴伐利亚州-德累斯顿动物园 - 长颈鹿园
4.德国柏林-Imchen square Berlin-Kladow Redesign of square and waterfront promenade
5，6.德国柏林-Imchen square Berlin-Kladow Redesign of square and waterfront promenade
7，8.德国柏林-国家展会（ULAP）广场

Pavement ○ 路面

1-4.德国柏林-晋城市儿童公园
5，6.德国柏林-莫阿比特监狱历史公园

Pavement ○ 路面

Pavement ○ 路面

1-3.德国慕尼黑-巴伐利亚国家博物馆
4.德国柏林-莫阿比特监狱历史公园
5,6.德国-什未林国家花园展
7,8.帝景天成坡地别墅

Pavement ○ 路面

1-3.帝景天成坡地别墅
4-8.东方普罗旺斯（欧式）

1，2.复地朗香
3.观塘别墅庭院
4.东莞市-湖景壹号庄园私家庭院
5，6.东山墅
7，8.都柏林-倾斜空间-gal 3

Pavement ○ 路面

105

Pavement ○ 路面

1-3.都柏林-诺曼底-NORMANDIE. Design by Hugh Ryan. Photo by H Ryan
4.俄勒冈州俄勒冈市-乔恩斯托姆公园
5,6.法国-Mente La-Menta景观设计
7,8.法国波尔多-Le miroir d'eau

1，2.法国克洛尔-Renewal of the main street
3，4.翡翠岛
5-9.佛山-天安鸿基花园

Pavement ○ 路面

109

Pavement ○ 路面

1. 广东东莞-东莞长城世家
2，3.广东佛山-山水庄园高档别墅花园
4，5.广东省-美的总部大楼
6.广州-保利国际广场
7.广州-汇景新城别墅

1-3.河源城区-河源东江·首府花园
4-6.荷兰Cuijk-Maaskade Cuijk
7，8.荷兰Getsewoud Zuid-schoolyard and playground Getsewoud Zuid

Pavement ○ 路面

113

Pavement ○ 路面

1. 荷兰海牙-square Beukplein
2，3.荷兰-解放公园
4，5.华盛顿-水桌庭园
6.汇景新城
7.霍恩博格海滨公园

Pavement ○ 路面

1，2.加利福尼亚－特伦顿大道
3.加利福尼亚 －加州威尼斯
4，5.加利福尼亚－特伦顿大道
6.加利福尼亚－马里布海岸豪宅
7.加利福尼亚圣塔莫妮卡-Euclid Park

Pavement ○ 路面

1，2.加拿大魁北克-Canadian Museum of Civilization Plaza
3，4.加拿大魁北克-Core Sample
5，6.加拿大蒙特利尔-Square Dorchester—Place du Canada

Pavement ○ 路面

1.加拿大蒙特利尔-Square Dorchester—Place du Canada
2.江南华府
3，4.江苏－睢宁流云水袖桥
5.江西省-爱丁堡二期
6.杰克·埃文斯船港—堤维德岬一期工程

Pavement ○ 路面

1-3. 金碧湖畔6号
4. 杰克·埃文斯船港—堤维德岬一期工程
5, 6. 曼萨纳雷斯线性公园
7, 8. 马萨诸塞州-波士顿儿童博物馆广场
9. 美国-哥伦比亚广场

1. 兰阿姆斯特丹-public square Columbusplein
2-4. 乐山市-恒邦·翡翠国际社区
5. 兰阿姆斯特丹-public square Columbusplein
6. 兰桥圣菲91号
7. 金碧湖畔6号

Pavement ○ 路面

125

Pavement ○ 路面

1.里昂市-里昂河岸
2.洛杉矶-南加州大学医学中心
3-5.绿洲千岛花园
6.美国波特兰-街区

Pavement 路面

1-3.美国马萨诸塞州波士顿-Levinson Plaza, Mission Park
4.美国加利福尼亚-特伦顿大道
5.美国旧金山-Cow Hollow学校操场

1,3.美国纽约-亚瑟·罗斯平台
2,4-6.美国普罗维登斯-The Steel Yard
7,8.梦归地中海

Pavement ○ 路面

1.梦归地中海
2.墨尔本-Elwood海滩景观设计
3.墨尔本-儿童艺术游乐园
4.泉州－奥林匹克花园别墅区67木栈道
5-7.泉州－奥林匹克花园别墅区

Pavement ○ 路面

1，2.南京-复地朗香别墅区
3.纽约-龙门广场州立公园
4-7.慕尼黑-施瓦宾花园城市
8，9.南海市民广场和千灯湖公园

Pavement ○ 路面

1. 纽约-龙门广场州立公园
2. 纽约-探索中心03-Dopress-CFD-ϑAlee
3. 纽约-威廉斯堡滨水演出会场
4，5.挪威-Gudbrandsjuvet-Viewing platforms && bridges
6. 挪威-Videseter railings
7. 挪威阿克斯胡斯-Kjenn市政公园，Lørenskog

Pavement ○ 路面

Pavement ○ 路面

1.挪威奥斯陆-Pilestredet公园
2,3.挪威阿克斯胡斯-Kjenn市政公园，Lørenskog
4,5.盘龙城

Pavement ○ 路面

1.萨缪尔·德·尚普兰滨水长廊
2.日式枯山水庭院－竹子栅栏
3，4.瑞士日内瓦－Renewal of Place des Nations
5.萨缪尔·德·尚普兰滨水长廊

1.桑德贝滨水景观，亚瑟王子码头
2，3.上海－宝绿园
4-6.上海－别墅花园
7.上海－－辰山植物园Climbing Plants2

Pavement ○ 路面

Pavement ○ 路面

1-3.上海－兰乔圣菲
4.上海－底楼花园
5.上海－丽茵别墅
6，7.上海－绿城
8，9.上海－绿洲江南

1.上海-玫瑰园
2.上海浦东-中邦城市雕塑花园
3,4.上海-仁恒屋顶花园

Pavement ○ 路面

1-3.上海-圣安德鲁斯庄园
4，5.，7上海市-比华利别墅
6.上海市-比利华1

Pavement ○ 路面

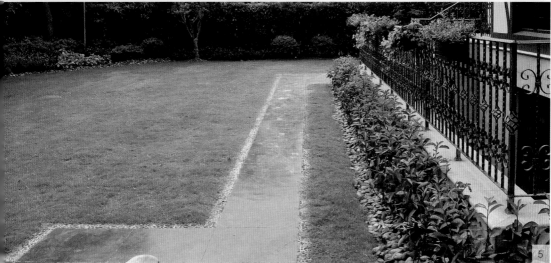

1，2.上海市-东町庭院
3.上海市-大华锦绣
4，5.上海市-观庭
6.上海市-豪嘉府邸别墅花园

Pavement ○ 路面

Pavement ○ 路面

上海市-平江路底楼花园
上海市-佘山高尔夫夜景
4.上海市-华庭雅居别墅花园
圣淘沙跨海步行道
世爵源墅
8.斯塔斯福特

1. 上海市-锦麟天地雅苑景观设计
2-4. 上海市 - 圣安德鲁斯庄园
5, 6. 上海市-太阳湖大花园设计造园

Pavement ○ 路面

Pavement ○ 路面

1-5.上海市－汤臣高尔夫别墅
6.上海市－万科燕南园

Pavement ○ 路面

1,2.上海市-新律花园
3.上海市-御翠园
4,5.上海市-月湖山庄假山水景
6.上海-太原路小别墅

1. 上海－太原路小别墅
2. 上海－汤臣
3，4. 佘山－东紫园
5，6. 上海－汤臣高尔夫
7-9. 上海－田园别墅花园

Pavement ○ 路面

Pavement ○ 路面

1-3.上海－万科朗润园
4-7.上海－万科深蓝

1，2.上海－西郊大公馆
3，4.上海－夏州花园
5-7.上海－香溪澜院
8.上海－小别墅
9.上海－新律花园

Pavement ○ 路面

Pavement ○ 路面

1-3.佘山－高尔夫别墅花园
4,5.深圳－万科金域华府
6-8.深圳－卧式摩天楼

Pavement ○ 路面

1，2.深圳-中海大山地
3，4.沈阳市-汇程庭院
5-7.佘山－东紫园
8，9.佘山－高尔夫别墅

Pavement ○ 路面

1.斯卡伯勒海滩城市设计规划1、4阶段
2，3.台湾－茂顺生活会馆
4-7.台湾－南投农舍别墅－waterfall rest area
8.泰国曼谷-Prive by Sansiri

Pavement ○ 路面

1. 唐山－唐人起居
2. 提香别墅
3，4. 提香草堂
5. 万成华府
6，7. 天津－桥园公园

173

Pavement ○ 路面

1.万科-金域华府
2.万科-兰乔圣菲
3,4.万科朗润园
5.万科棠樾
6.威尔克斯巴里堤防加高河流公地
7.威尼斯
8.威尼斯水城
9.威尼斯水城别墅花园

Pavement ○ 路面

，2.小金家
.新港码头
.新加坡-City Square Urban Park
.新南威尔士帕丁顿-PADDINGTON RESERVOIR
GARDENS
.威特市斯博拉茨
，8.文华别墅
.五溪御龙湾

Pavement ○ 路面

.五栋大楼屋顶花园 - 屋顶花园
.武汉 - 天下别墅
.西班牙圣塞瓦斯蒂安-AITZ TOKI 别墅
.西班牙伊伦市-Alai Txoko公园
-7.泰国曼谷-Prive by Sansiri

Pavement ○ 路面

1.西雅图－山顶住宅
2.悉尼-Pirrama公园
3，4.西雅图－山顶住宅
5.悉尼-Pirrama公园

1.悉尼-Pirrama公园
2-4.悉尼-奥林匹克公园Jacaranda广场
5.悉尼-海洋生物站公园
6-8.香碧歌庄园

Pavement ○ 路面

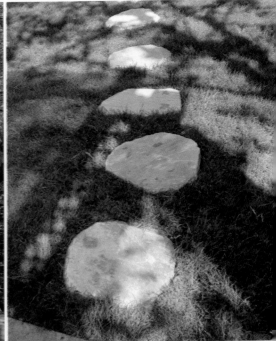

Pavement ○ 路面

1，3.亚利桑那州-理工学院景观
2.亚德韦谢姆大屠杀博物馆大楼
4.雪梨澳乡别墅庭院
5.雅居乐私家花园

1,2.亚利桑那州-梅萨艺术中心
3,4.阳光九九造园
5.阳光闲庭1
6.以色列贝尔谢巴-BGU University Entrance Square && Art Gallery
7.以色列哈尔阿达尔-Har Adar #2住宅
8.依云听香阁后花园

Pavement ○ 路面

187

1，2.易郡
3，4.意大利-Via Regina公共花园
5.意大利博洛尼亚-CIRCOLARE, ECOLE DEL RUSCO
6.意大利卢卡-圣·安娜区公园
7.意大利洛迪-漫步花园

Pavement ○ 路面

Pavement 路面

1. 钟楼海滩
2. 重庆－梦中天地
3-5. 竹怡居
6，7. 棕榈湾

1. 棕榈湾
2, 3. 云栖蝶谷
4. 浙江省-浮动花园-永宁河公园

GARDEN LANDSCAPE DETAIL DESIGN · RAMP
坡道

　　庭院的意义在于居住者对它所提供的生活方式的认同，在于它所带来的院落"能指"（居住环境）和"所指"（居住者的生活方式、社会交往方式）的和谐关系，而此代码在现代集合住宅中的被扭曲和遗忘引发了居民对院落中亲密的邻里关系的怀念。

　　庭院作为人为的自然空间，它将自然和人工相结合起来，让人产生一种回归自然的向往。庭院中往往常用小巧、精致的手法，设置植物、院路、亭廊、假山、雕塑、水池等，既表现出个性化的特点，同时给人们带来自然美、人工美的艺术享受。

gnatus eiumendae voluptamusa custet et demporeribus minverchil mi, unte magnit eum et que doloreh enihili aspe vendi ium nume sum vita doloria nosam as utendi te solori beribus pa consequi idias maiorem porestiis erendias as es ut volupit, el earum labo. Itatemp orerenes ipsunt, simusap idustenet explacea et. on rat doluptatus arum et que plit occum et iunt quam, voluptia parcius dollect aturehentum abo. Neque voluptas eque sam, qui opta volorib ernate lat latem eumet la quis qui ut arit, am reperates dolore.

1.Australia-Adelaide Zoo Entrance Precinct_
2.Aalborg, Denmark-奥尔堡滨水 - 港口与城市连接
3-5.Barcelona - 蒙锥克公园的新缆车车站
6.Belo Horizonte, Brazil Date of Completion-Pampulha广场
7.Brumadinho, Brazil Date of Completion-Burle Marx教育中心
8.Canada-cSimcoe WaveDeck

Ramp ○ 坡道

Ramp ○ 坡道

1. Canada-cSimcoe WaveDeck
2. Celle Ligure－眺望台及铁路改造
3. Fredericia - Temporary Park
4. Linz, Austria-乡村别墅公园和散步广场
5，6. Shreveport－雨花园别墅
7. Sitges－肯·罗伯特公园2110 13 copia
8. Taiwan, China Date of Completion-日月潭观光局管理处
9. 澳大利亚墨尔本-墨尔本717号Bourke大街

Ramp ○ 坡道

1-4.澳大利亚墨尔本-墨尔本717号Bourke大街
5.北京市海淀区-新诗意山居

1，2.德国 柏林-南"制革厂"岛
3.韩国首尔-West Seoul Lake Park
4，5.荷兰Cuijk-Maaskade Cuijk
6.澳大利亚悉尼-Ballast Point 公园
7，8.荷兰阿姆斯特丹-Floriande居住区

Ramp ○ 坡道

Ramp ○ 坡道

1. 荷兰阿姆斯特丹-potgieterstraat
2-4. 荷兰海牙-playground Melis Stokepark
5. 加拿大-东南福溪可持续社区
6. 江苏-睢宁流云水袖桥

1，2.墨尔本-儿童艺术游乐园
3.江西省-爱丁堡二期
4.绿洲千岛花园7678
5.曼萨纳雷斯线性公园cpr3©anamuller_300
6-8.墨尔本-儿童艺术游乐园

Ramp ○ 坡道

1.深圳-SHENZHEN BAY COASTLINE PARK
2.深圳－卧式摩天楼IMG_1756-med
3，4.圣淘沙跨海步行道
5-8.威尔克斯巴里堤防加高河流公地
9.西安－世园会之荷兰生态园－new_holland garden

Ramp ○ 坡道

1.以色列贝尔谢巴-BGU University Entrance Square && Art Gallery
2.意大利博洛尼亚-CIRCOLARE, ECOLE DEL RUSCO
3.澳大利亚墨尔本-墨尔本国际会议会展中心
4.澳大利亚墨尔本-墨尔本717号Bourke大街

GARDEN LANDSCAPE DETAIL DESIGN · STEPS

台阶

　　庭院的意义在于居住者对它所提供的生活方式的认同，在于它所带来的院落"能指"（居住环境）和"所指"（居住者的生活方式、社会交往方式）的和谐关系，而此代码在现代集合住宅中的被扭曲和遗忘引发了居民对院落中亲密的邻里关系的怀念。

　　庭院作为人为的自然空间，它将自然和人工相结合起来，让人产生一种回归自然的向往。庭院中往往常采用小巧、精致的手法，设置植物、院路、亭廊、假山、雕塑、水池等，既表现出个性化的特点，同时给人们带来自然美、人工美的艺术享受。

Agnatus eiumendae voluptamusa custet et demporeribus minverchil mi, unte magnit eum et que doloreh enihili quaspe vendi ium nume sum vita doloria nosam as utendi te solori beribus pa consequi idias maiorem porestiis exerendias as es ut volupit, el earum labo. Itatemp orerenes ipsunt, simusap idustenet explacea et. on rat doluptatus arum et que plit occum et iunt quam, voluptia parcius dollect aturehentum abo. Neque voluptas eque sam, qui opta volorib ernate lat latem eumet la quis qui ut arit, am reperates dolore.

1-4.Aalborg, Denmark 奥尔堡滨水 港口与城市连接
5._上海－圣安德鲁斯庄园
6.Aley, Lebanon Date of Completion-消失点
7.Anchorage, Alaska, USA-Anchorage Museum Expansion

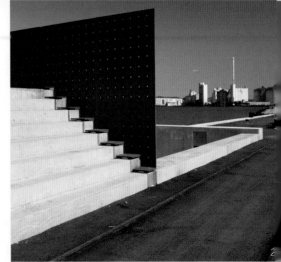

Steps ○ 台阶

Steps ○ 台阶

1.Belo Horizonte, Brazil Date of Completion-Pampulha广场
2.Bassil Mountain Escape, Faqra, Lebanon
3.Barcelona-蒙锥克公园的新缆车车站-TELEFERICO_20_VISTA
4.Barcelona, Spain Date of Completion-Tanatorio Ronda de Dalt花园
5，6.Austin, USA Date of Completion-串联式溪边住宅

213

Steps 〇 台阶

.Berlin－三角铁路站公园－Com_Loidl_BerlinPARK_20110904-0268 Kopie
.Connecticut Country House, Westport, Connecticut
.Brumadinho, Brazil Date of Completion-Burle Marx教育中心
-6.Connecticut Country House, Westport, Connecticut
.Faqra－魔力居所FEJ_9056

Steps ○ 台阶

1. Faqra - 魔力居所
2. FZK_022
3. Greenwich Residence, Greenwich, Connecticut
4. Hilltop Residence, Seattle, WA
5. Genk Belgium-m!ne

1-4, 6.Holstebro, Denmark-If it is outstanding, celebrate it twice
5.Los Angeles, CA-Wilmington Waterfront Park
7. Horizon Residence, Venice, California

Steps 〇 台阶

1，2.House by the Creek, Dallas, Texas
3.Malibu Beach House, Malibu, California
4.Malinalco House, Malinalco, State of Mexico, Mexico
5.PA-323-09
6.Private ResidenceGarden of Planes, Richmond, VA
7.Quartz Mountain Residence, Paradise Valley, Arizona

1-3.Unfolding Terrace, Dumbo, Brooklyn, New York
4.Speckman House Landscape, Highland Park, St. Paul, MN
5.Sitges－肯·罗伯特公园2110 42 copia
6，7.Urban Play Garden
8.Villengarten Krantz

Steps ○ 台阶

223

Steps ○ 台阶

爱尔兰都柏林-登陆
, 3.Wittock Residence
Villengarten Krantz
, 6.Zapopan-七庭院住宅 = 7 Patios House
澳大利亚墨尔本-墨尔本717号Bourke大街

1-5.澳大利亚悉尼-Ballast Point 公园

Steps ○ 台阶

Steps ○ 台阶

1-4.澳大利亚悉尼-Bondi 至Bronte 滨海走廊
5.澳大利亚悉尼-Victoria Park Public Domain

1.鲍恩海滨
2,3.芭堤雅希尔顿酒店
4,5.百家湖
6.北京-波特兰
7.北京-翠湖——教授的乐园

Steps ○ 台阶

1. 北卡罗来纳州阿什维尔-xueNorth 卡罗莱纳州植物
2. 北京-燕西台
3，4.北悉尼地区-BP公司遗址公园
5，6.比利时Kapellen-Meeting square Kapellen

Steps ○ 台阶

Steps ○ 台阶

1. 成都－翡翠城汇锦云天5
2. 成都－玉都别墅庭园
3. 德国 柏林-南"制革厂"岛
4. 德国巴伐利亚02007 Waldkirchen花园展
5. 德国柏林-Imchen square Berlin-Kladow Redesign of square and waterfront promenade
6，7.德国柏林-国家展会（ULAP）广场

1. Melboume－圣基尔达海滨城的交通线
2. Malinalco House, Malinalco, State of Mexico, Mexico
3. Manhattan Roof Terrace, New York, New York
4. Melboume－圣基尔达海滨城的交通线
5，6.帝景天成坡地别墅
7.东莞市-湖景壹号庄园私家庭院

Steps ○ 台阶

1.广东东莞-东莞长城世家
2,3.广东佛山-山水庄园高档别墅花园
4.广州-凤凰城8号
5.韩国汉城-ChonGae Canal Restoration Project
6.韩国首尔-West Seoul Lake Park

Steps ○ 台阶

Steps ○ 台阶

1. 韩国首尔-West Seoul Lake Park
2，3.荷兰Cuijk-Maaskade Cuijk
4.荷兰Velsen-public park Duinpark
5.华南碧桂园燕园

241

1.荷兰Vught-Parklaan Zorgpark Voorburg
2.加拿大-私人公寓，兰乔圣菲
3.加利福尼亚-马里布海岸豪宅
4-6.加拿大-东南福溪可持续社区
7.加拿大魁北克-Canadian Museum of Civilization Plaza

Steps ○ 台阶

243

1. 江南华府2
2. 佛山－天安鸿基花园
3. 金碧湖畔6号
4-7. 杰克·埃文斯船港—堤维德岬—期工程

Steps ○ 台阶

Steps ○ 台阶

乐山市-恒邦·翡翠国际社区
君临5栋
绿洲千岛花园3
里昂市-里昂河岸
洛杉矶 - 南加州大学医学中心RCH_LAC USC_
onner_5489
芬洛市-马斯河大街—芬洛马斯河岸新城区

1.绿洲千岛花园99
2.绿洲千岛花园DPP_35309
3，4.美国加利福尼亚-PV庄园
5，6.美国纽约-Carnegie Hill House

Steps ○ 台阶

1，2.美国纽约-亚瑟·罗斯平台
3.美国普罗维登斯-The Steel Yard
4.墨尔本 - Frankston - DSCF0838
5-7.墨尔本-儿童艺术游乐园

Steps ○ 台阶

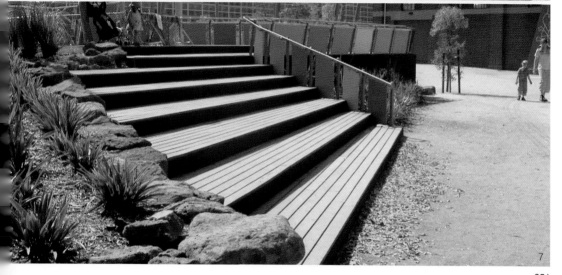

Steps ○ 台阶

1.墨尔本-儿童艺术游乐园
2.墨西哥－马里纳尔可住宅
3.慕尼黑-施瓦宾花园城市
4，5.南海市民广场和千灯湖公园
6.南京－复地朗香别墅区

1，2.挪威奥斯陆-Pilestredet公园
3，4.挪威奥斯陆-Rolfsbukta住宅区
5，6.挪威奥斯陆-Tjuvholmen

Steps ○ 台阶

2.泉州－奥林匹克花园别墅区－调整大小 泉奥0618 038
4.萨缪尔·德·尚普兰滨水长廊
桑德贝滨水景观，亚瑟王子码头
上海－绿洲江南

257

Steps ○ 台阶

1-3.上海－绿洲江南
4，5.上海－玫瑰园
6.上海浦东-中邦城市雕塑花园

Steps ○ 台阶

1. 上海浦东－中邦城市雕塑花园
2. 上海市－比华利别墅
3. 上海－圣安德鲁斯庄园
4. 上海市－大华锦绣
5，6.上海市－圣安德鲁斯庄园

1.上海市－圣安德鲁斯庄园
3，4.上海市－汤臣高尔夫别墅
5-7.上海－万科深蓝
8.上海市－御翠园
9.上海－银都名墅别墅花园

Steps ○ 台阶

1.圣淘沙跨海步行道
2.深圳－卧式摩天楼
3-5.斯卡伯勒海滩城市设计规划1、4阶段
6.台湾－南投农舍别墅－backyard
7.斯塔斯福特

Steps ○ 台阶

1，2.斯塔斯福特
3，4.泰国曼谷-Prive by Sansiri
5-7.特拉福德滨河长廊

Steps ○ 台阶

1. 威尔克斯巴里堤防加高河流公地
2，3. 西郊大公馆107号庭院
4. 土耳其-雅勒卡瓦克海滨码头
5. 西郊一品
6. 西雅图-山顶住宅
7，8. 悉尼-Pirrama公园

Steps ○ 台阶

Steps ○ 台阶

1，2.悉尼-海洋生物站公园-CC-02
3，4.香碧歌庄园
5，6.新南威尔士帕丁顿-PADDINGTON RESERVOIR GARDENS
7，8.新西兰-杰利科北部码头漫步长廊，杰利科大道和筒仓公园

1.以色列-海尔兹利亚公园
2,3.银都名墅
4.云间绿地
5.云间绿大地
6.云栖蝶谷
7.浙江金华-艾青文化公园

Steps ○ 台阶

273

1.浙江金华-义乌江大坝
2.浙江金华-艾青文化公园
3.中国杭州-山湖印别墅
4.中山市-中山别墅花园
5.重庆－梦中天地
6.尊

Steps ○ 台阶

1.Unfolding Terrace, Dumbo, Brooklyn, New York
2.柏林-Südkreuz火车站
3，4.钟楼海滩

GARDEN LANDSCAPE DETAIL DESIGN · ENCLOSING WALL

围墙

  庭院的意义在于居住者对它所提供的生活方式的认同，在于它所带来的院落"能指"（居住环境）和"所指"（居住者的生活方式、社会交往方式）的和谐关系，而此代码在现代集合住宅中的被扭曲和遗忘引发了居民对院落中亲密的邻里关系的怀念。

  庭院作为人为的自然空间，它将自然和人工相结合起来，让人产生一种回归自然的向往。庭院中往往常采用小巧、精致的手法，设置植物、院路、亭廊、假山、雕塑、水池等，既表现出个性化的特点，同时给人们带来自然美、人工美的艺术享受。

gnatus eiumendae voluptamusa custet et demporeribus minverchil mi, unte magnit eum et que doloreh enihili uaspe vendi ium nume sum vita doloria nosam as utendi te solori beribus pa consequi idias maiorem porestiis xerendias as es ut volupit, el earum labo. Itatemp orerenes ipsunt, simusap idustenet explacea et. on rat doluptatus arum et que plit occum et iunt quam, voluptia parcius dollect aturehentum abo. Neque voluptas eque sam, qui opta volorib ernate lat latem eumet la quis qui ut arit, am reperates dolore.

Enclosing wall ○ 围墙

2.8a The Terrace
Aalborg, Denmark-奥尔堡滨水 - 港口与城市连接
Austin, USA Date of Completion-串联式溪边住宅
6.Australia-Adelaide Zoo Entrance Precinct

Enclosing wall ○ 围墙

Il nuovo Parco di Jean Nouvel a Barcellona
Lausanne - 绿窗_L011_Myriam Ramel 4
Greenwich Residence, Greenwich, Connecticut
5.Il nuovo Parco di Jean Nouvel a Barcellona
罗德岛普罗维登斯 - Dunkin Donuts Plaza -
Horizon Garden
7.Lakeway Drive - In Redevlopment

Enclosing wall ○ 围墙

Lausanne－绿窗_L011_Myriam Ramel 5
3.Lee Landscape, Calistoga, CA
London－韦尔斯利大道和公园路－OKRA_
oydon_pers_03
Madrid RIO
Malibu Beach House, Malibu, California
8.Malibu Beach House, Malibu, California

1.Oss, the Netherlands Date of Completion-库肯霍夫街区
2，3.Pamet Valley
4.Private ResidenceGarden of Planes, Richmond, VA
5，6.Salvokop, Pretoria, Gauteng-The Freedom Park
7.爱尔兰都柏林-登陆

Enclosing wall ○ 围墙

Enclosing wall ○ 围墙

1，2.Speckman House Landscape, Highland Park, St. Paul, MN
3.阿尔布兰茨瓦尔德的Delta精神病治疗中心
4. Woody Creek Garden, Pitkin County, Colorado
5.爱尔兰都柏林-Cleansweep
6，7.爱尔兰都柏林-凡丹戈
8.澳大利亚-东联高速公路

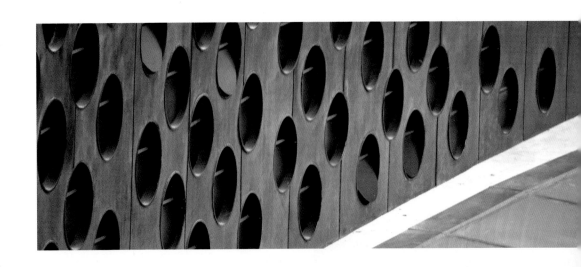

Enclosing wall ○ 围墙

1，2.芭堤雅希尔顿酒店
3，4.澳大利亚堪培拉国家紧急救灾服务纪念馆
5.澳大利亚悉尼-Ballast Point 公园

Enclosing wall ○ 围墙

1.澳大利亚堪培拉国家紧急救灾服务纪念馆
2，3.澳大利亚悉尼-Ballast Point 公园
4.澳大利亚悉尼-Ng别墅花园
5.芭堤雅希尔顿酒店
6.鲍恩海滨
7.白香果
8.北京-低碳住家——北京褐石公寓改造设计
9.北京-燕西台

Enclosing wall ○ 围墙

1. 碧桂园
2. 北京-新新家园
3. 北京-易郡别墅
4. 碧湖山庄
5. 成都-翡翠城汇锦云天
6. 城中豪宅
7. 德国巴伐利亚州-德累斯顿动物园-长颈鹿园
8. 德国巴伐利亚州-伯格豪森的巴伐利亚州园林展-城市公园

Enclosing wall ○ 围墙

1，2.德国柏林-晋城市儿童公园
3，4.德国-什未林国家花园展
5.德国巴伐利亚州-德累斯顿动物园-长颈鹿园

1，2.帝景天成坡地别墅
3.东山墅
4，5.佛山－天安鸿基花园
6，7.广东东莞-东莞长城世家

Enclosing wall ○ 围墙

1，2.芬洛市-马斯河大街—芬洛马斯河岸新城区
3，4.广东省-美的总部大楼
5.荷兰阿姆斯特丹-Floriande居住区
6.韩国首尔-West Seoul Lake Park
7.荷兰海牙-playground Melis Stokepark

Enclosing wall ○ 围墙

Enclosing wall ○ 围墙

1. 旧金山－私人住宅
2. 康涅狄格州郊区住宅
3. 华盛顿特区－国家动物园公园亚洲径
4. 加拿大，私人公寓，兰乔圣菲
5，6. 兰桥圣菲91号

1. 洛杉矶－南加州大学医学中心RCH_LAC USC_Bonner_5489_011C
2. 曼萨纳雷斯线性公园arg3© anamuller_300
3. 泉州－奥林匹克花园别墅区67－休闲平台
4. 美国纽约－Carnegie Hill House
5. 明尼阿波利斯－史拜克曼住宅景观，319_11
6. 梦归地中海
7. 挪威奥斯陆-Rolfsbukta住宅区
8，9. 挪威-Gudbrandsjuvet -Viewing platforms & bridges

Enclosing wall ○ 围墙

Enclosing wall ○ 围墙

1. 上海－万科深蓝
2. 上海－万科燕南园
3. 上海－西郊大公馆
4. 上海市－汤臣高尔夫别墅
5. 上海－新律花园
6. 上海－银都名墅别墅花园

1.上海－御翠园
2.斯卡伯勒海滩城市设计规划1、4阶段
3.佘山－高尔夫别墅花园
4.深圳－万科金域华府－清幽的竹林
5.沈阳－万科新里程2.

1.斯塔斯福特
2.台湾－南投农舍别墅－waterfall rest area
3.台湾－茂顺生活会馆-03
4.台湾商会
5.万科－兰乔圣菲
6.泰国曼谷-Prive by Sansiri

Enclosing wall ○ 围墙

1. 泰国曼谷-Prive by Sansiri
2，3. 同润加州
4. 土耳其-雅勒卡瓦克海滨码头
5-7. 万科棠樾

Enclosing wall ○ 围墙

1.西郊大公馆107号庭院
2.西郊花园
3.万科棠樾
4.西郊一品
5.纽约-伸展的平台
6.悉尼-奥林匹克公园Jacaranda广场
7.新加坡-City Square Urban Park

Enclosing wall ○ 围墙

1，2.新加坡-City Square Urban Park
3.新南威尔士帕丁顿-PADDINGTON RESERVOIR GARDENS
4.亚利桑那州-梅萨艺术中心207-03
5.新南威尔士帕丁顿-PADDINGTON RESERVOIR GARDENS
6.新西兰-杰利科北部码头漫步长廊，杰利科大道和筒仓公园

1. 亚德韦谢姆大屠杀博物馆大楼
2. 亚利桑那州－梅萨艺术中心207-05
3. 亚利桑那州－石英山居所，534-07
4. 亚洲大酒店
5. 阳光九九造园－009
6. 阳光闲庭11

Enclosing wall ○ 围墙

Enclosing wall ○ 围墙

3.以色列哈尔阿达尔-Har Adar #2住宅
5.易郡20090708007
银都名墅

1.意大利-Via Regina公共花园
2.万科-金域华府